About the book

There are 12 GCE A Level Mathematics papers and answers in this book. These are 4 sets of papers 1, 2 & 3 written as practice papers for GCE Mathematics Examinations in June 2021. Papers are mainly focusing on Edexcel, AQA, OCR & OCR MEI examination boards. However, you may still use this book as a practice for other GCE A Level examination boards as well.

These papers are written according to the new 2017 syllabus and questions are potential questions for the upcoming examinations in June 2021.

All the questions in this book are written by the author and they are new questions written purely to help and experience the students to prepare and test themselves for the upcoming A Level mathematics exam.

There are 4 sections to this book A, B, C & D. Each section contains 3 papers. All 3 papers of each section are calculator papers.

A Level Mathematics

June 2021

Potential Papers

Includes answers

for the Edexcel, OCR, AQA & OCR MEI 2017 syllabuses

(can also be used as a revision guide for other exam boards)

By Dilan Wimalasena

Contents

Section A

GCE A Level Mathematics

Pure Mathematics 1

Potential Paper 1A

June 2021

Students must have Mathematical Formulae
and Statistical Tables, Calculator.

Calculator is allowed

**Time allowed
2 hours
Total 100 marks**

Write answers to 3 significant figures unless stated
otherwise

1. Show that

$$\sqrt{\tan^3 x \sec x \, cosec^3 x} = 1 + \tan^2 x$$

(3 marks)

2. A curve C has equation

$$y = 2x^3 - \frac{7}{2}x^2 + 2x + 1$$

Work out the coordinates of stationary points of C & determine their nature.

(8 marks)

3. OPQ is a sector of a circle. OB = 4cm, BQ = 2cm and arc AB = 3cm. Work out the perimeter & area of the shaded region.

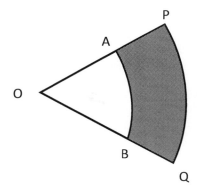

(5 marks)

4. $f(x) = 6x^3 + x^2 - 9x - 4.$
   i)   Factorise $f(x)$ completely.

(4 marks)

   ii)   Sketch the curve $y = f(x)$.

(3 marks)
(total 7 marks)

5. a) Find the fist three terms in ascending powers of x, of the binomial expansion of
$$\sqrt{9 - 2x}$$

Giving each coefficient in its simplest form.

(4 marks)

b) By substituting a suitable value for x to your expansion in part (a), find an approximate value for $\sqrt{8.98}$ to 3 decimal places.

(3 marks)
(total 7 marks)

6. $f(x) = kx^2 - kx + 5, k \neq 0.$

   i) $f(x) = 0$ has only one real root. Work out the value of k.

(3 marks)

   ii)   Given that $k > 0$, sketch the graph of $y = f(x)$.

(2 marks)
(total 5 marks)

7. Solve the equation for $-2\pi \leq \theta \leq 2\pi$

$$\frac{\cos4\theta + \cos2\theta}{\sin4\theta - \sin2\theta} = \sqrt{2}, \theta \neq n\pi$$

(7 marks)

8. The price of a car can be modelled by the formula

$$P = 550 + 24000e^{-\frac{t}{4}}$$

Where, P is the price of the car in £s and t is the age of the car in years after being purchased.

a) Calculate the new price of the car.

(2 marks)

b) Calculate the price after 5 years.

(2 marks)

c) When will it be worth less than £5000?

(3 marks)

d) Work out the price of the car as $t \rightarrow \infty$.

(2 marks)

e) Sketch the graph showing P against t.
   Comment on the appropriateness of this model.

(3 marks)
(total 12 marks)

9. $f(x) = 3x - x^2$

    i)   Sketch the curve $y = f(x)$.

(2 marks)

    ii)   The normal to the curve at the origin meets the curve again at the point P.

        a) Work out the equation of the normal to the curve at the origin.

(4 marks)

        b) Hence, or otherwise work out the coordinates of the point P.

(3 marks)

    iii)   Work out the area of the region bounded by the curve and the normal in part (ii).

(4 marks)

(total 13 marks)

10.i) Sketch on the same axes the graphs of $y = |x - 3|$ & $y = 5 - |4x|$, clearly showing coordinates of any intersections with the coordinate axes.

(5 marks)

ii) Hence, or otherwise solve $|x - 3| = 5 - |4x|$

(4 marks)
(total 9 marks)

11. Find the area of the finite region bounded by the curve with parametric equations

$$x = 2t^3, y = \frac{4}{t}, t \neq 0.$$

the x axis, the values $x = 16$ & $x = 54$.

(6 marks)

12. A curve C has equation

$$4x^2 - 3y^2 - 3x - 7xy + 5 = 0$$

Find the equation of the tangent to C at the point $(1, -3)$ giving your answer in the form $ax + by + c = 0$, where a, b & c are integers.

(7 marks)

13. Prove that

$$S_\infty = \frac{a}{1 - r}$$

for any geometric series with $|r| < 1$.

(4 marks)

14. a) Sketch the curves $y = -3 + \ln 2x$ & $y = \frac{2}{x}$ on the same axes.

(2 marks)

b) Two curves intersect at $x = a$.
Show that

$$a - 2 + a \ln 2a = 0$$

(1 marks)

c) Using the iterative formula

$$x_{n+1} = \frac{1}{2} e^{\frac{2}{x_n}+3}, x_0 = 12$$

Find a to 3 significant figures.

(4 marks)
(total 7 marks)

Total for paper: 100 marks

End

GCE A Level Mathematics

Pure Mathematics 2

Potential Paper 2A

June 2021

Students must have Mathematical Formulae
and Statistical Tables, Calculator.

Calculator is allowed

Time allowed
2 hours
Total 100 marks

Write answers to 3 significant figures unless stated
otherwise

1. The function f is defined by

$$f: x \rightarrow \frac{3x + 4}{x - 2}, \{x \in \mathbb{R}, x > 2\}$$

   i.   Find the $f^{-1}(x)$

   (2 marks)

   ii. Find the range of $f^{-1}(x)$

   (1 mark)

   iii. Find the domain of $f^{-1}(x)$

   (2 marks)

   (total 5 marks)

2. In the diagram below, $\overrightarrow{OA} = a, \overrightarrow{OB} = b, OC:CA = 2:1, AD:DB = x:y$.
   Given that OBE is a straight line and $\overrightarrow{OE} = 2\overrightarrow{OB}$.

   Find the values of x & y.

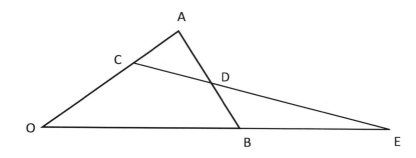

   (5 marks)

3. Figure below shows a sector of a circle.

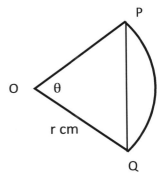

i.   Prove that the area of the segment is
    $\frac{1}{2}r^2(\theta - \sin\theta)$, where θ is in radians.

(3 marks)

ii. Distance PQ is 4cm and $r = 6cm$.
    Work out the area of the segment.

(3 marks)
(total 6 marks)

4. $y = \sin^2\theta$, $x = \cos\theta - 3, 0 \le \theta \le \frac{\pi}{2}$

i)   Work out $\frac{dy}{dx}$ in terms of θ.

(3 marks)

ii) Work out the equation of tangent at $x = -\frac{5}{2}$.

(4 marks)
(total 7 marks)

5. $f(x) = x^3 - x^2 - 3x + 3, x \in \mathbb{R}.$
   a. Find $f(1)$

   (1 mark)
   b. Hence, or otherwise write $f(x)$ as a product of two algebraic factors.

   (2 marks)
   c. Using part (b) or otherwise, solve

   $$\cot^3 \theta - \cot^2 \theta - 3 \cot \theta + 3 = 0 \text{ for } -\pi \leq \theta \leq \pi$$

   (4 marks)
   (total 7 marks)

6. Solve the following
   a) $\operatorname{cosec} \theta + 2 = \sin \theta + \cot \theta \, (3 + \cos \theta)$
      in the interval $0 \leq \theta \leq 2\pi$, giving your answers to 3 significant figures.

   (6 marks)
   b) $\tan^2 x - 3 = \sec x \, (2 - \sec x)$
      in the interval $0 \leq x \leq 360°$, giving your answers to 1 decimal place.

   (6 marks)
   (total 12 marks)

7. Find
   i) $\sum_{r=1}^{12} 10(3^r)$

   (3 marks)

   ii) the sum to infinity of the geometric series
   $$\frac{2}{9} + \frac{1}{6} + \frac{1}{8} + \cdots$$

   (3 marks)
   (total 6 marks)

8. Prove that
   $$\frac{d(\cot x)}{dx} = -\text{cosec}^2 x$$

   (4 marks)

9. $f(\theta) = 24 \sin \theta - 7 \cos \theta$.
   Given that $f(\theta) = R \sin(\theta - \alpha)$, where $R > 0 \ \& \ 0 \le \alpha \le \frac{\pi}{2}$
   i) Find the values of $R \ \& \ \alpha$.

   (4 marks)

   ii) Hence, solve the equation
   $$24 \sin \theta - 7 \cos \theta = 13 \text{ for } 0 \le \theta \le 2\pi$$

   (5 marks)

   iii) a. Find the greatest value of $(24 \sin \theta - 7 \cos \theta)^3$.

   (2 marks)

   b. What is the smallest positive value of θ when the greatest value in part iii) a. occurs.

   (3 marks)
   (total 14 marks)

22

10. Showing your method clearly in each case, find

   i) $\int 6\tan^2 x \sec^2 x \, dx$

   (5 marks)

   ii) $\int_1^2 x^2 \ln x \, dx$

   (5 marks)
   (total 10 marks)

11. A curve has equation
$$xe^y - ye^x = 1$$

   i.   Find in terms of x & y an expression for $\frac{dy}{dx}$.

   (4 marks)

   ii.   Show that the equation of tangent to the curve at $(0, -1)$ is
$$ey - ex + e - x = 0$$

   (4 marks)
   (total 8 marks)

12.

   a) Express $\frac{3}{9-y^2}$ in partial fractions.

   (3 marks)

   b) Hence, obtain the solution of
$$3\tan x \frac{dy}{dx} = (9 - y^2)$$

   For which $y = 1$ at $x = \frac{\pi}{6}$, giving your answer in the form

$$\sin^2 x = g(y)$$

   (8 marks)
   (total 11 marks)

13. Given that

$$x = \operatorname{cosec}\frac{y}{3}, 0 < y < \pi$$

Show that

$$\frac{dy}{dx} = \frac{-3}{x\sqrt{x^2 - 1}}$$

(5 marks)

Total for paper: 100 marks

End

GCE A Level Mathematics

Statistics and Mechanics

Potential Paper 3A

June 2021

Students must have Mathematical Formulae
and Statistical Tables, Calculator.

Calculator is allowed

Time allowed
2 hours
Total 100 marks

Write answers to 3 significant figures unless stated
otherwise

SECTION A: STATISTICS

1. A variable X is normally distributed.

   Given that $P(x < 19) = 0.35$ & $P(x > 37) = 0.45$.

   i)   Find the mean and standard deviation of X.

   (6 marks)

   ii)   Calculate the value of a, where $P(x > a) = 0.7$.

   (4 marks)
   (total 10 marks)

2. A bag contains 4 red sweets, 5 yellow sweets and 3 green sweets. Holly ate two sweets.

   i)   Draw a tree diagram to represent all possible outcomes for her two sweets.

   (3 marks)

   ii)   Work out the probability of her eating the same coloured sweets.

   (3 marks)

   Given that she has picked two different coloured sweets.

   iii)   Work out the probability that her second sweet is red.

   (4 marks)
   (total 10 marks)

3. A company receives a mean of 150 incoming calls a day. However, the manager of the company believes that the mean number of calls is higher than 150 calls.

   The manger conducts an experiment and collects the following data. Where, x is the number of calls and n is the number of days.

   $$n = 60, \sum x = 9360, \sum x^2 = 1464086$$

   i.   Find the mean and the standard deviation of the sample.

   (4 marks)

   ii.   Carry out a hypothesis test to test the managers' claim at the 1% significance level. State your assumptions clearly.

   (5 marks)
   (total 9 marks)

4. In a study of how much time footballers spend using the treadmill, the data for a random sample of 11 players are examined below.

   $$23, 41, 19, 38, 32, 24, 25, 18, 44, 34, 29$$

   i.   Work out the mean and the standard deviation for the data.

   (3 marks)

   ii.   Work out the median and the quartiles for the data.

   (3 marks)

   iii.   Show that there are no outliers.

   (2 marks)

   iv.   Draw a box plot for these data.

   (3 marks)
   (total 11 marks)

5. A company produces wine glasses in large quantities. It is known from previous records that 1% of the glasses have a hairline fracture.

A random sample of 250 glasses was taken from the production line.

   i.   Calculate the probability that exactly 5 glasses have a hairline fracture, using

         a) Binomial distribution

         b) Normal distribution

                                                            (3 marks)

In a separate experiment, a random sample of 300 glasses were taken and 10 of them had hairline fractures. The inspector claims, that the probability of a glass with a hairline fracture has changed.

   ii.   Using a normal approximation, test at the 2% level of significance, whether the inspectors claim is correct.

                                                            (7 marks)

Total for section A is 50 marks

SECTION B: MECHANICS

6. A particle of mass 5 kg rests on a rough horizontal table. Particle A is attached to a light inextensible string which passes over a smooth pulley fixed to the edge of the table. The other end of the string is connected to particle B of mass 8 kg which hangs freely below the pulley 1.2m above the ground.

   The system is released from rest with the string taut. Particle A does not reach the pulley before B reaches the ground. The coefficient of friction between the particle A and the table is 0.15.

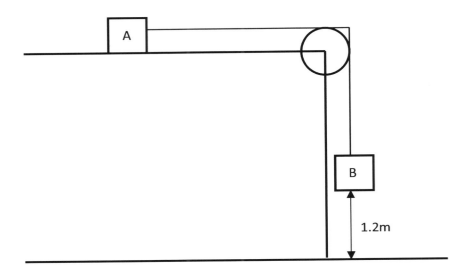

i.   Calculate the acceleration of the system before B hits the ground.

(3 marks)

ii.   Calculate the tension before B hits the ground.

(2 marks)

iii. Calculate the total time taken by A to come to rest from the beginning.

(5 marks)
(total 10 marks)

7. A particle is projected from a height of 10m above the ground with speed $24ms^{-1}$ at an angle α to the horizontal, where $\tan \alpha = \frac{12}{5}$.

i.   Calculate the greatest height reached above the ground by the particle.

(3 marks)

ii.   Calculate the horizontal distance travelled before reaching the ground.

(6 marks)

iii.   Work out the speed of the particle when it is 6m above the ground level.

(5 marks)
(total 14 marks)

8. A ladder PQ, of mass 2kg and length 6a, has one end P resting on rough horizontal ground. The other end Q rests against a smooth vertical wall. A load of mass 5kg is fixed on the ladder at the point S, where PS = 2a. The ladder is modelled as a uniform rod in a vertical plane perpendicular to the wall and the load is modelled as a particle. The ladder rests in limiting equilibrium making an angle of 30° with the wall.

i)   Draw a diagram showing all the forces acting on the ladder.

(1 mark)

ii)   Find the coefficient of friction between the ladder and the ground.

(10 marks)
(11 marks)

9. A plank XY has mass 50 kg and length 5 m. A load of mass 15 kg is attached to the plank at Y. The loaded plank is held in equilibrium, with XY horizontal, by two vertical ropes attached at X and Z. The plank is modelled as a uniform rod and the load as a particle. Given that the tension in the rope at Z is two times the tension in the rope at X.

    Calculate
    (a) the tension in the rope at Z.

    (2 marks)

    (b) the distance ZY.

    (5 marks)
    (total 7 marks)

10. A box of mass 7.5 kg is pulled up a rough plane by means of a light rope. The plane is inclined at an angle of 25° to the horizontal. The rope is parallel to a line of greatest slope of the plane. The tension in the rope is 65 N. The coefficient of friction between the box and the plane is 0.2. By modelling the box as a particle,

    find
    (a) the normal reaction of the plane on the box.

    (3 marks)

    (b) the acceleration of the box.

    (5 marks)
    (total 8 marks)

Total for section B is 50 marks

Total for paper: 100 marks

End

Section B

GCE A Level Mathematics

Pure Mathematics 1

Potential Paper 1B

June 2021

Students must have Mathematical Formulae
and Statistical Tables, Calculator.

Calculator is allowed

Time allowed
2 hours
Total 100 marks

Write answers to 3 significant figures unless stated
otherwise

1. $f(x) = 12x^3 - 19x^2 - 13x + 6$

Factorise $f(x)$ completely

(4 marks)

2. Given that $cos\theta = \sqrt{2} - 1$

a) Find the value of $cos2\theta$ in the form $a + b\sqrt{2}$, where a & b are integers.

(3 marks)

Given that $2\cos(y + 45) = \sqrt{2}\sin(y - 45)$

b) Show that $tany = 1$

(5 marks)
(total 8 marks)

3. $f(x) = \dfrac{x^2+1}{3x-4}, \ x \in \mathbb{R}, x \neq \dfrac{4}{3}$

    a) Find and simplify an expression for $f'(x)$

(3 marks)

    b) Find the set of values of x for which $f(x)$ is decreasing.

(5 marks)
(total 8 marks)

4. The first four terms in the series expansion of $(1 + ax)^n$ in ascending powers of x are

$$1 + \frac{2}{3}x + \frac{8}{9}x^2 + kx^3 + \cdots$$

Where a, n & k are constants and $|ax| < 1$.

i)   Find the values of a & n?

(4 marks)

ii)   Show that $k = \dfrac{112}{81}$

(2 marks)
(total 6 marks)

5. $f(x) = 3x^2 - 12x + 7$

a) Write $f(x)$ in the form $p(x + q)^2 + r$
Where p, q & r are integers to be found.

(3 marks)

b) Sketch $y = f(x)$ showing any points of intersection with coordinate axes and coordinates of the turning point.

(3 marks)

c) Describe fully the transformation that maps the curve $y = f(x)$ onto the curve $y = g(x)$ where

$$g(x) = 3(x - 1)^2 - 12x + 12$$

(3 marks)

d) Find the range of the function

$$h(x) = \frac{10}{3x^2 - 12x + 7}$$

(1 mark)
(total 10 marks)

6. i.) Find the exact value of x such that

$$3 \arcsec (x + 1) - \pi = 0$$

(3 marks)

ii.) Solve, for $0 < \theta < 2\pi$, the equation

$$\sin 2\theta + 2 \sin \theta - \cos \theta = 1$$

Giving your answers in terms of π.

(5 marks)
(total 8 marks)

7. i.) Find, as natural logarithms, the solutions of the equation

$$e^{2x} - 19e^x + 48 = 0$$

(3 marks)

ii.) Use proof by contradiction to prove that $\log_3 5$ is irrational.

(5 marks)
(total 8 marks)

8. The curve C has equation $y = (x + 1)(x - 3)^2$

    i) Sketch the curve, clearly showing coordinates of any intersections with coordinate axes.

    (3 marks)

    ii) Work out the equation of the tangent to curve at $x = 4$.

    (3 marks)

    iii) Find the area of the region bounded by the curve and the tangent above the x-axis.

    (5 marks)
    (total 11 marks)

9. i.) Find the value of (a) such that

    $$\log_a 32 = 5 + \log_a 243$$

    (3 marks)

    ii.) Solve the equation

    $$3^{x+2} = 5^{x-1}$$

    Giving your answer to 3 significant figures.

    (4 marks)
    (total 7 marks)

10. i.) Sketch on the same axes the graphs of

$y = 9a^2 - x^2$ & $y = |3x - a|$ where a is a positive constant.

Show, in terms of a, the coordinates of any points where each graph meets the coordinate axes.

(4 marks)

ii.) Find the exact solutions of the equation

$$9 - x^2 = |3x - 1|$$

(4 marks)
(total 8 marks)

11. In year 1990, the population of a town was 35,000. A model of the growth of the population assumes that (n) years after 1990, the population (P) is given by

$$P = 35000 \times 1.03^n$$

Using this model

a) Find the population in 2010?

(2 marks)

b) Find, to the nearest day, the time when the initial population have doubled.

(4 marks)
(total 6 marks)

12. $y = x^2 e^{-x}$

     i)   Find coordinates of stationary points of $y = x^2 e^{-x}$.

                         (5 marks)

     ii)   Find the equation of normal to $y = x^2 e^{-x}$ at $x = 1$.

                         (3 marks)
                   (total 8 marks)

13. Given that

$$f(x) = \frac{(x-3)(x-1)}{(3x-1)(1+x)^2} \, , |x| > \frac{1}{3}$$

    a. Express $f(x)$ in partial fractions.

                         (3 marks)

    b. Show that

$$\int_1^2 f(x)\, dx = p[\ln q + r]$$

    Where p, q & r are rational numbers.

                         (5 marks)
                   (total 8 marks)

Total for paper: 100 marks

End

GCE A Level Mathematics

Pure Mathematics 2

Potential Paper 2B

June 2021

Students must have Mathematical Formulae
and Statistical Tables, Calculator.

Calculator is allowed

Time allowed
2 hours
Total 100 marks

Write answers to 3 significant figures unless stated
otherwise

1. i) Express $\sqrt{28.9}$ in the form $k\sqrt{10}$

   where k is a rational number.

   (2 marks)

   ii) Find integer n such that

   $$2\sqrt{50} - \sqrt{32} = \sqrt{n}$$

   (3 marks)
   (total 5 marks)

2. Figure below shows the curve with equation
   $$y = \sqrt{5x - 2}$$

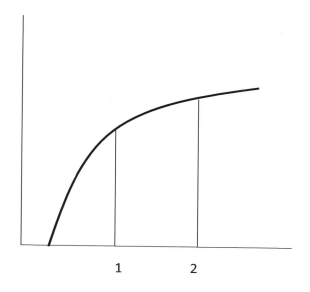

Use the trapezium rule with 4 equally spaced ordinates to estimate the area of the region between the curve, the x-axis and the lines $x = 1$ & $x = 2$.

(4 marks)

3. Sector OPQ below has a perimeter $26cm$ & arc PQ is $10cm$.

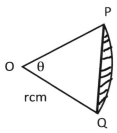

i) Find r & θ

(2 marks)

ii) Find the area of the shaded region?

(3 marks)
(total 5 marks)

4. A curve has parametric equations

$$x = cosec\,\theta - \cot\theta\ \&\ y = \sec\theta - \tan\theta$$

i) Show that

$$x + \frac{1}{x} = 2\cosec\theta$$

(3 marks)

Given that $y + \frac{1}{y} = 2\sec\theta$

ii) Find cartesian equation of the curve.

(3 marks)

iii) Find an expression for $\frac{dy}{dx}$ in terms of x & y.

(4 marks)
(total 10 marks)

5. i)   Use the substitution $u = 3 - x^3$ to find

$$\int \frac{x^2}{3 - x^3} \, dx$$

(3 marks)

ii)   Evaluate

$$\int_0^{\frac{\pi}{4}} \sin 3x \cos x \, dx$$

(5 marks)
(total 8 marks)

6. The function f is defined by

$$f(x) = 2e^{x-2}, x \in \mathbb{R}$$

a) State the range of f.

(1 mark)

b) Find an expression for $f^{-1}(x)$ and state its domain

(3 marks)

   The function g is defined by $g(x) = 3x - 1, x \in \mathbb{R}$

   Find in terms of e

c) The value of $gf(\ln 3)$

(3 marks)

d) The solution to the equation

$$f^{-1}g(x) = 5$$

(3 marks)
(total 10 marks)

7. A vendor is selling goods at a market stall. She believes that the rate at which she makes sales depends on the length of time since the start of the sale, t hours, and the total value of sales she has made up to that time, £x.

She uses the model

$$\frac{dx}{dt} = \frac{k(4-t)}{x}, where\ k\ is\ a\ constant$$

Given that after 4 hours she has made sales of £84 in total.

i)   Solve the differential equation and show that she has make £56 in the first hour. (to nearest pound)

(6 marks)

The vendor believes that it is not worth staying at the stall once she is making sales at a rate less than £12 per hour.

ii)   Verify that 4 hours and 2 minutes after the start of the sale, she should have already left.

(4 marks)
(total 10 marks)

8. The 2nd & 3rd terms of a geometric series are $\log_2 9$ & $\log_2 81$ respectively.

   a) Find common ratio.

   (3 marks)

   b) Show that the first term $a = \log_2 3$

   (2 marks)

   c) Find to 2 decimal places, the sum of the first 7 terms of the series.

   (3 marks)
   (total 8 marks)

9. Points A, B & C are such that $A(1, 0, 2), B(2, 5, -2)$ & $C(4, -1, 2)$

   i)   Find the vectors $\overrightarrow{AB}, \overrightarrow{AC}$ & $\overrightarrow{BC}$

   (3 marks)

   ii)   Find $\left|\overrightarrow{AB}\right|, \left|\overrightarrow{AC}\right|$ & $\left|\overrightarrow{BC}\right|$

   (2 marks)

   iii) Work out the angle BAC

   (2 marks)

   iv)   Hence, or otherwise find the area of triangle ABC

   (2 marks)
   (total 10 marks)

10. Figure shows the graph of $y = f(x)$ which meets the coordinate axes at the points $(a, 0)$ & $(0, b)$, where a and b are constants.

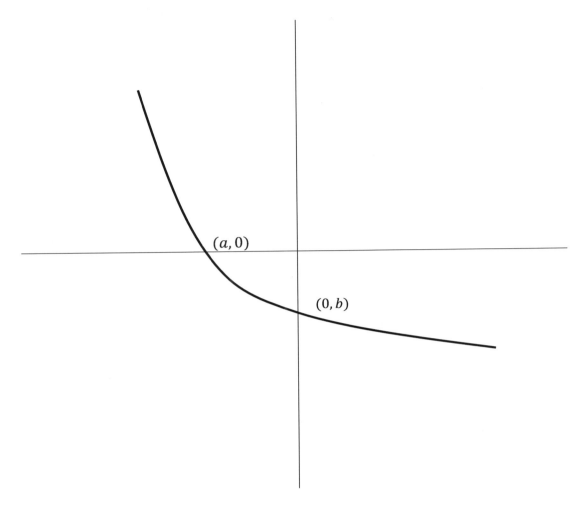

i)   Showing, in terms of a & b, the coordinates of any points of intersections with the axes, sketch on separate diagrams the graphs of

   a) $y = f^{-1}(x)$ b) $y = 3f(2x)$

   (4 marks)

   Given that $f(x) = 1 - \sqrt{x + 4}, x \in \mathbb{R}, x \geq -4$

ii)   Find a & b.

                                                                    (2 marks)

iii) Find an expression for $f^{-1}(x)$ and state its domain.

                                                                    (4 marks)
                                                                    (total 10 marks)

11. $f(x) = \sin x - \ln x, x > 0$

   a) Show that $f(x) = 0$ has a root in the interval $[\,2.2, 2.3\,]$

                                                         (2 marks)

   b) Find $f'(x)$

                                                         (2 marks)

   c) Using $x_0 = 2.2$ as a first approximation, apply the Newton-Raphson procedure to find a second approximation, giving your answer to three decimal places.

                                                         (3 marks)
                                                    (total 7 marks)

12. $f(\theta) = 24 \sin \theta - 7 \cos \theta$

   Given that $f(\theta) = R\sin(\theta - \alpha)$, where $R > 0$ and $0 \le \alpha \le \frac{\pi}{2}$

   a) Find the value of R & α. (α to 3 decimal places)

                                                         (4 marks)

   b) Hence, solve for $0 \le \theta \le 2\pi$, the equation

$$24 \sin \theta - 7 \cos \theta = 10$$

                                                         (3 marks)

   c) Calculate the maximum value of

$$h(\theta) = \frac{15}{30 - 24 \sin 3x + 7 \cos 3x}$$

                                                         (3 marks)

   d) Find the smallest positive value of x for which this maximum occurs.

                                                         (3 marks)
                                                  (total 13 marks)

Total for paper: 100 marks

End

GCE A Level Mathematics

Statistics and Mechanics

Potential Paper 3B

June 2021

Students must have Mathematical Formulae
and Statistical Tables, Calculator.

Calculator is allowed

**Time allowed
2 hours
Total 100 marks**

Write answers to 3 significant figures unless stated
otherwise

SECTION A: STATISTICS

1. The probability function of a discrete rando variable x is given by

$$p(x) = kx^2, x = 1,2,3$$

Where x is a positive constant.

a. Show that $k = \frac{1}{14}$

(2 marks)

b. Find $p(x > 1)$

(2 marks)

c. Find $E(x)$

(2 marks)

d. Find $Var(2 - x)$

(4 marks)
(total 10 marks)

2. Marks scored in a mathematics test by 60 students is summarised below

Marks	Frequency
40-49	15
50-69	12
70-79	18
80-89	13
90-100	2

i)   Use linear interpolation to estimate the median mark

(2 marks)

ii)   Estimate the mean mark

(2 marks)

iii)   Calculate the standard deviation

(3 marks)

iv)   Estimate the interquartile range

(3 marks)
(total 10 marks)

3. At the start of a competition there are 6 contestants of which 4 are female. In each round of the competition, one contestant is eliminated. All contestants have the same chance of going through to the next round.

   a. Show that the chance of first 3 contestants to be eliminated are female is $\frac{1}{5}$.

   (2 marks)

   b. Find the probability there are more females than males are eliminated in the first 3 rounds.

   (5 marks)

   c. Given that the first contestant to be eliminated is female, find the probability that the next 2 are female.

   (3 marks)
   (total 10 mark)

4. Jay's Sweets claims that 1 in 8 packets contains a free gift.

   Jay's Sweets can be bought in a family pack containing 8 packets. Find the probability the packets in one of the family packs contain

   a) No gifts

   (2 marks)

   b) More than 3 gifts

   (2 marks)

   Jay's Sweets can also be bought wholesale in boxes containing 40 packets. A school canteen noticed that their students only found 2 gifts in the packets from one of these boxes.

   c) Stating your hypothesis clearly, test at the 10% level of significance whether or not this gives evidence of there being fewer gifts than it is claimed in the advert.

   (4 marks)
   (total 8 marks)

5. In a hospital, 20% of patients are over 90 years old and 60% are under 60 years old.

    i)   Find the mean age and standard deviation of ages.

                                      (3 marks)

    ii)   There were 180 patients in the hospital on a particular day. Estimate number of patients between 50-80 years old.

                                        (3 marks)

                                   (total 6 marks)

6.

    a. State the conditions under which the binomial distribution can be approximated by normal distribution.

                                        (2 marks)

$$X \sim B(25, 0.3)$$

    b. Use binomial distribution to work out $P(X < 9)$

                                        (1 mark)

    c. Use normal distribution to work out $P(X < 9)$

                                        (1 mark)

    d. Calculate the percentage error between parts (b) & (c).

                                        (2 marks)

                                        (total 6 marks)

Total for section A is 50 marks

SECTION B: MECHANICS

7. At time t seconds, where $t \geq 0$, a particle P moves such that its velocity, $v\ ms^{-1}$, is given by

$$v = (4 - 9t^2)i - (4t)j$$

When $t = 2$, the displacement of the particle from fixed origin O is $(i + j)m$.

Find the distance of the particle from O when $t = 5$ seconds, giving your answer to 3 significant figures.

(9 marks)

8. A projectile is fired with velocity $35\ ms^{-1}$ at an angle of elevation 45^0 from a point P on horizontal ground. The projectile moves freely under gravity until it reaches the ground at the point Q. Find:

   a. The distance PQ

(5 marks)

   b. The speed of the projectile at the point when it is $7m$ above the ground.

(5 marks)
(total 10 marks)

9. A particle of mass $0.4kg$ is being pulled up a rough slope that is angled at 20^0 to the horizontal by a force of $4N$. The force acts on an angle 35^0 to the slope.

   Given that the coefficient of friction is 0.09.

   i)   Calculate the acceleration of the particle.

                                                              (7 marks)

   The $4N$ force is now removed.

   ii)   Calculate the time taken for the particle to return to the original position which was 5m.

                                                              (6 marks)
                                                         (total 13 marks)

10. A ladder AB of mass 30kg and length 8m, rests with its base, A, on a rough horizontal ground and it top, B, leaning against a smooth vertical wall. The coefficient of friction between the ladder and the ground is 0.35. the ladder lies in a perpendicular plane vertical to the wall and the ground and is inclined at angle 50^0 to the horizontal.

   A weight of 65kg is hung to the ladder. Modelling the weight as a particle and the ladder as a uniform rod, find the maximum distance up the ladder the weight can be hung before the ladder begins to slip.

                                                              (10 marks)

11. Particles A & B of masses 3m kg and m kg are attached to the ends of a light inextensible string whish passes over a smooth pulley. The both hang at a distance of 2.5m above horizontal ground. The system is released from rest.

   a. Find the magnitude of the acceleration of the particles

   (3 marks)

   b. Work out the speed of A as it hits the ground.

   (2 marks)

   Given that B does not reach the pulley.

   c. Find the greatest height B reaches above the ground.

   (3 marks)
   (total 8 marks)

Total for section B is 50 marks

Total for paper: 100 marks

End

Section C

GCE A Level Mathematics

Pure Mathematics 1

Potential Paper 1C

June 2021

Students must have Mathematical Formulae
and Statistical Tables, Calculator.

Calculator is allowed

**Time allowed
2 hours
Total 100 marks**

Write answers to 3 significant figures unless stated
otherwise

1. Andrew saves money as follows. £5 on day 1, £6 on day 2, £7 on day 3 and so on.

    (i)   Work out the amount he would have saved after 7 days.

    (2 marks)

    (ii)   How many days will it take him to save £1000?

    (3marks)

    (total 5 marks)

2. AB is a diameter of a circle centre C. Where A (1, 5) & B (3, 9)

    (i)   Work out the coordinates of C.

    (1 marks)

    (ii)   Write down equation of circle.

    (2 marks)

    (iii)   The tangent to this circle at B, meets x-axis at P and y-axis at Q. Calculate the distance PQ.

    (5 marks)

    (total 8 marks)

3. $y = x^2 - 8x + 3$

    (i)   Find $\frac{dy}{dx}$

(1 marks)

    (ii)   Find $\int y\,dx$

(2 marks)

    (iii)   Work out equation of normal to y at x = -2.

(3 marks)
(total 6 marks)

4. a) $\dfrac{2}{(2-x)(1+x)^2} = \dfrac{A}{(2-x)} + \dfrac{B}{(1+x)} + \dfrac{C}{(1+x)^2}$

    Find A, B & C.

(4 marks)

    b) Hence expand,

    $\dfrac{2}{(2-x)(1+x)^2}$ in ascending powers of x as far as x^2 term.

(6 marks)

(total 10 marks)

5. Points A, B & C are as follows

$$A \begin{pmatrix} 1 \\ 2 \\ 3 \end{pmatrix}, B \begin{pmatrix} 5 \\ 0 \\ 6 \end{pmatrix}, C \begin{pmatrix} 9 \\ -2 \\ 6 \end{pmatrix}$$

(i)   Work out vectors $\overrightarrow{AB}$ & $\overrightarrow{AC}$

(3 marks)

(ii)   Calculate angle BAC

(4 marks)

(iii)   Hence, work out area of triangle ABC.

(3 marks)

(total 10 mark)

6. Sector OPQ below has angle POQ $= 1.5^c$ & Arc PQ $= 4.5cm$

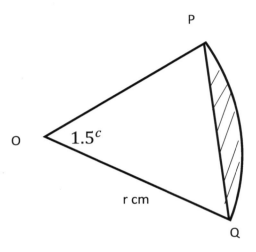

i)   Find r.

(2 marks)

ii)   Find area of shaded region.

(4 marks)

(total 6 marks)

7. Prove that if $y = x^2$ then $\frac{dy}{dx} = 2x$ using first principles of differentiation.

(4 marks)

8. A population is growing according to the following formula

$$P = 15e^{2.5t}, t \geq 0$$

$$t = years, P = population\ in\ thousands$$

i)   Find the initial population

(2 marks)

ii)   Find the time taken for the population to grow from double the initial population to triple the initial population.

(4 marks)

(total 6 marks)

9. Given that $x = cosec^2 2y$, $0 < y < \frac{\pi}{4}$

    i)   Find $\frac{dx}{dy}$

(3 marks)

    ii)   Show that $\frac{dy}{dx} = \frac{-1}{4x(x-1)^{\frac{1}{2}}}$

(4 marks)

    iii)   Find $\frac{d^2y}{dx^2}$ in terms of x. Give your answer in its simplest form.

(4 marks)

(total 11 marks)

10. $x = 2 - cot\theta$, $y = 1 - cosec\theta$

    i)   Find y in terms of x.

(3 marks)

    ii)   Find $\frac{dy}{dx}$ when $\theta = \frac{\pi}{3}$

(4 marks)

    iii)   Find normal to curve at $\theta = \frac{\pi}{3}$

(4 marks)

(total 11 marks)

11. Function f is such that

$$f(x) = 2 - \frac{1}{x+1} - \frac{2}{(x+1)^2}, \qquad x \neq -1$$

i)   Show that

$$f(x) = \frac{2x^2 + 3x - 1}{(x+1)^2}, \qquad x \neq -1$$

(4 marks)

ii)   Show that $2x^2 + 3x - 1 \geq c$ for a constant c.

Write down the constant c.

(3 marks)

iii)   Find $f'(x)$

(3 marks)

(total 10 marks)

12. $f(\theta) = 7\cos\theta - 3\sin\theta$

  i)   Write $f(\theta)$ in the form $R\cos(\theta + \alpha)$

         where $R > 0$ & $0 < \alpha < \dfrac{\pi}{2}$

         Find R & α to 3 significant figures.

                                                                                (4 marks)

  ii)   Show that

         $5\sec\theta + 3\tan\theta = 7$ can be written as $7\cos\theta - 3\sin\theta = 5$

                                                                                (3 marks)

  iii)   Hence or otherwise, solve for $0 \le x \le \pi$

         $$5\sec 2x + 3\tan 2x = 7$$

         Write your solutions to 3 significant figures.

                                                                                (6 marks)

                                                                         (total 13 marks)

_____

                                                       Total for paper: 100 marks

                                   End

GCE A Level Mathematics

Pure Mathematics 2

Potential Paper 2C

June 2021

Students must have Mathematical Formulae
and Statistical Tables, Calculator.

Calculator is allowed

Time allowed
2 hours
Total 100 marks

Write answers to 3 significant figures unless stated
otherwise

1. Find

$$\int 56x^{5/2} \, dx$$

(3 marks)

2. $3, \; x + 5, \; 7x + 5$ are 3 consecutive terms of a geometric sequence. Find x.

(3 marks)

3. Expand the following up to x^3 term in ascending powers of x.

$$\frac{1}{(3 - 4x)^3} \, , \, |x| < \frac{3}{4}$$

(5 marks)

4. The functions f & g are defined by

$$f(x) = 2e^{x-3}, x \in \mathbb{R}$$
$$g(x) = 3x - 1, x \in \mathbb{R}$$

i)   State the range of f

(1 marks)

ii)   Find $f^{-1}(x)$ and state its domain

(3 marks)

iii) Find the exact value of $gf(ln4)$

(3 marks)

iv)   Solve $f^{-1}g(x) = 5$

(3 marks)
(total10 marks)

5.
   i)   Simplify

$$\frac{x^3 - 8}{3x^2 - 8x + 4}$$

(4 marks)

   ii)   Solve $4e^{2x} - 8e^x + 3$
         Write your answers in exact form

(4 marks)
(total 8 marks)

6. Differentiate with respect to x

   i) $y = 3\cos^2 x + cosec\,2x$

   (3 marks)

   ii) $y = (x + \ln(3x))^2$

   (3 marks)

   iii) $y = \dfrac{3x^2 - 8x + 7}{(x-2)^2}$

   (4 marks)
   (total 10 marks)

7. A pond weed growth in a lake is monitored. Initially there were (A) pondweeds in the lake.

   A suggested model for (n) thousand pondweeds after (t) months is

   $$n = \dfrac{600}{5 + 7e^{\frac{-t}{4}}}$$

   i)   Find the value of A

   (2 marks)

   ii)   Find the number of months it takes to reach 75 thousand pondweeds in the lake.

   (4 marks)

   iii)   Model suggest that the pondweeds cannot exceed N thousand. Find N.

   (3 marks)
   (total 9 marks)

8.

    i)   Prove that

$$cot\theta - tan\theta \equiv 2cot2\theta, \qquad \theta \neq \frac{n\pi}{2}$$

(3 marks)

    ii)   Solve for $-\pi < \theta < \pi$, the equation

$$cot\theta - tan\theta = 5 \text{ (to 3 significant figures)}$$

(6 marks)
(total 9 marks)

9.

    i)   Sketch $y = |3x|$ and $y = |x - 2|$, on the same axes.

(4 marks)

    ii)   Solve $|3x| = |x - 2|$

(4 marks)

    iii)   Solve $2 - x = 2|x + 1|$

(4 marks)
(total 12 marks)

10. Work out the following integrals

i) $\int_0^{\frac{1}{3}} \frac{e^{3x}}{2e^{3x}-1} dx$

(5 marks)

ii) $\int_0^{\frac{\pi}{6}} \tan^2 x \, dx$

(5 marks)

iii) $\int_0^{\frac{\pi}{3}} \cos^2 x \, dx$

(5 marks)
(total 15 marks)

11. The volume of a spherical air bubble of radius r cm is V cm^3, where $V = \frac{4}{3}\pi r^3$.

i) Work out
$$\frac{dV}{dr}$$

(2 marks)

ii) The volume of the bubble increases with time t seconds according to formula

$$\frac{dV}{dt} = \frac{500}{(4t + 1)^2} , t \geq 0$$

Work out
$$\frac{dr}{dt}$$

(3 marks)

iii)   Given that $V = 0$, *when* $t = 0$.
        Solve the differential equation above to obtain V in terms of t.

(6 marks)

iv)   Hence, at time $t = 10$,

   a) Find the radius of the bubble to 3 significant figures.

(2 marks)

   b) Find rate of increase of the radius of the bubble.

(3 marks)
(total 16 marks)

Total for paper: 100 marks

End

83

GCE A Level Mathematics

Statistics and Mechanics

Potential Paper 3C

June 2021

Students must have Mathematical Formulae
and Statistical Tables, Calculator.

Calculator is allowed

Time allowed
2 hours
Total 100 marks

Write answers to 3 significant figures unless stated
otherwise

SECTION A: STATISTICS

1. During a football tournament mean number of goals scored by a team during a match is 2.1 goals with a standard deviation of 0.5 goals.

    i)   Suggest a suitable probability distribution for number of goals scored by a team during a match.

    (1 mark)

    ii) Using your suggested distribution to part (i), work out percentage of times a team has scored more than 3 goals in a match.

    (3 marks)

    iii) Work out probability of a team scoring between 1.5 goals and 2.5 goals in a match.

    (5 marks)

    iv) 80% of the time, teams have scored more than (x) goals in a match. Find the value of (x).

    (4 marks)
    (total 13 marks)

2. Average weekly time spent on revision by 6 students during weekends and their mock exam marks for a maths test are recorded below.

Student	A	B	C	D	E	F
Time Spent (minutes)	35	80	75	58	100	95
Marks	47	96	60	59	88	91

i) Above data can be coded using T (time spent) & M (marks) using the following formulae. Work out x & y values.

$$x = \frac{T - 50}{10} , y = \frac{M + 40}{100}$$

(2 marks)

ii) Calculate the standard deviation for x & y.

(4 marks)

iii) Write down the standard deviation for T & M.

(2 mark)

The regression equation for y & x is given below
$$y = 0.1 + 0.44x$$

iv) Work out equation of regression line for M on T.

(2 marks)

v) A student wants to find an estimate for marks he can expect to achieve if he is to spend 150 minutes of revision time. What is his estimate,

(2 mark)

vi) How reliable is his estimate in part (iv)?

(1 mark)
(total 13 marks)

3. Two events P & Q are such that

$$P(P) = 0.5, P(Q) = 0.42, P(P' \cup Q') = 0.86$$

Work out the following

i) $P(P \cap Q)$

(2 marks)

ii) $P(P \cup Q)'$

(2 marks)

iii) $P(P' \mid Q)$

(3 marks)

iv) $P(Q' \mid P)$

(3 marks)
(total 10 marks)

4. Information of some heights are given below.

Height (cm)	120-150	150-170	170-180	180-200
Frequency	8	14	11	7

i)   Represent above data on a histogram.

(3 marks)

ii)   Work out the mean and standard deviation for above information.

(3 marks)
(total 6 marks)

5. A school claims that 45% of its sixth form students achieve at least 1 grade A. A random sample of 20 students are chosen.

   i)   Assuming the schools claim is correct, calculate the probability of at least 9 students in the sample achieving grade A in one or more subjects.

                            (2 marks)

       Another sample of 80 students are taken.

   ii)   Write down the conditions under which the normal distribution can be used as an approximation to binomial distribution.

                            (1 mark)

   iii)   In this sample, 27 students achieved at least one grade A in their subjects. Test at 5% level of significance, whether the schools claim is correct.

                            (5 marks)
                        (total 8 marks)

Total for section A is 50 marks

SECTION B: MECHANICS

6. A truck of mass 1500kg is towing a trailer of mass 500kg. They are connected by a light inextensible tow bar, which is parallel to direction of motion. The truck engine produces a driving force of 4500N. The resistances to motion are constant and is 2R N for the truck and R N for the trailer.

   Given that the acceleration of the truck and trailer is $0.75ms^{-2}$.

   i)   Work out the value R.

   (4 marks)

   ii)   Find the tension in the tow bar.

   (3 marks)
   (total 7 marks)

7. Particle P accelerates from rest from point A with acceleration $2ms^{-2}$ and particle Q accelerates from rest 3 seconds after P from point A with acceleration $3ms^{-2}$.

   Calculate the distance from A where Q overtakes P?

   (total 6 marks)

8. A stone is projected vertically upwards with speed $U ms^{-1}$ from point A, which is 10m above sea level. The stone takes 6.75 seconds to drop down to sea level from projection.

    i)   Find the value U.

(3 marks)

    ii)   Work out the total distance travelled by the stone before hitting the sea level.

(3 marks)

    iii)   Find the time difference between the two points, where the stone is at least 11m above sea level.

(6 marks)
(total 12 marks)

9. Two particles A (3kg) & B (5kg) are attached to a smooth pulley by a light inextensible cable. Both particles are 2.5m above the ground as shown.

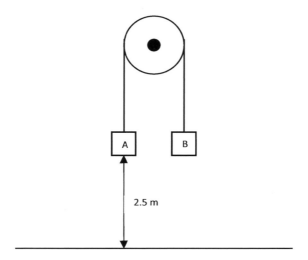

2.5 m

Once released, B hits ground and is immediately brought to rest. A does not reach the pulley in the subsequent motion.

i)   Calculate the force exerted on the pulley by two particles before B hits the ground.

(4 marks)

ii)   Calculate the total distance travelled by particle A, after release and until the cable becomes taught again.

(5 marks)

When the experiment was carried out at a different location, it was noticed that particle B had penetrated 3cm into the ground before coming to rest.

i)   Calculate the force exerted on B by the ground.

(6 marks)
(15 marks)

10. Particle A is resting on a rough plane inclined at an angle α to the horizontal, where $\tan \alpha = \dfrac{1}{\sqrt{3}}$ and the coefficient of friction between the plane and the particle is 0.3.

Particle A is held in equilibrium by a horizontal force P as shown.

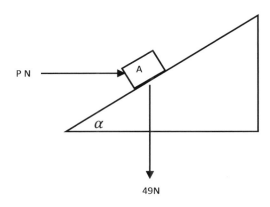

    i)   Work out the minimum and maximum possible values of P.

                                          (6 marks)

        Given that P is 100 N.

    ii)   Calculate the acceleration of A.

                                          (4 marks)
                               (total 10 marks)

Total for section B is 50 marks

Total for paper: 100 marks

End

Section D

GCE A Level Mathematics

Pure Mathematics 1

Potential Paper 1D

June 2021

Students must have Mathematical Formulae
and Statistical Tables, Calculator.

Calculator is allowed

Time allowed
2 hours
Total 100 marks

Write answers to 3 significant figures unless stated
otherwise

1. $y = x^2 - 10x - 3$

    i)   Sketch the above curve, clearly stating any intersections with axes and turning points.

                  (3 marks)

    ii)   Find equation of normal to the curve at $x = 2$.

                  (4 marks)

    iii)   Above normal to the curve intersects x-axis at A and y-axis at B. Calculate the area of triangle OAB where O is the origin.

                  (3 marks)

                  (total 10 marks)

2. A sector has radius r & angle θ.

    i)   Prove that the arc length is $r\theta$

                  (2 marks)

    ii)   Prove that the area of sector is $\frac{1}{2}r^2\theta$

                  (3 marks)

                  (total 5 marks)

3. A (2, 7), B (10, 3)
   AB is a diameter of a circle centred at C.

    i)   Find equation of circle

                  (3 marks)

    ii)   Find equation of tangent to circle at A.

                  (4 marks)

                  (total 7 marks)

4. Work out the following integrals

   i) $\int_0^{\frac{1}{2}} \dfrac{e^{2x}}{5e^{2x}+3}\,dx$

(4 marks)

   ii) $\int_{\frac{\pi}{6}}^{\frac{\pi}{4}} \cot^2 x \, dx$

(4 marks)

   iii) $\int_0^{\frac{\pi}{2}} 6\sin^2 x \, dx$

(5 marks)
(13 marks)

5. Prove that

$$sec\theta - tan\theta = \frac{2cos\theta - sin2\theta}{1 + cos2\theta}, \theta \neq \frac{n\pi}{2}$$

(4 marks)

6. The table below shows a company's growth in revenue (R).

Year	2005	2006	2007	2008	2009
Revenue (£1000's)	58.1	67.5	80.1	105.6	138.6

The data can be represented by an exponential model for revenue growth. Using t as number of years since the year 2000 & R as revenue in £1000's.

A suitable model is

$$R = a \times 10^{kt}$$

i)   Derive an equation for $\log_{10} R$ in terms of a, k & t.

(2 marks)

ii)   Draw the graph of R against t, using a best fit line.

(3 marks)

iii)   Using your line, work out R in terms of t.

(4 marks)

iv)   According to the model, calculate the revenue in year 2000.

(1 mark)

v)   According to the model, when will the revenue reach £250,000?

(2 marks)
(total 12 marks)

7. A geometric progression has 5th term 6075 & 8th term 225.

i)   Work out the first term and the common ratio.

(4 marks)

ii)   Find the difference between the sum to infinity and the sum of the first 10 terms.

(4 marks)
(total 8 marks)

8. $f(x) = x^3 + 5x^2 - 8x + 1$

   Work out the values of x, for which the function $f(x)$ is increasing.

(4 marks)

9. $x = 2\cot\theta$, $y = 3\sqrt{3}\,\text{cosec}\,\theta$, $0 < \theta < \dfrac{\pi}{2}$

The point P lies on the curve C with above parametric equations and at P, $\theta = \dfrac{\pi}{3}$.

i)   Work out the exact value of $\dfrac{dy}{dx}$ at P.

(5 marks)

ii)   Q also lies on C, where $\dfrac{dy}{dx} = 1$. Find exact coordinates of Q.

(4 marks)

iii)   Work out tangent to C at P.

(4 marks)
(total 13 marks)

10. a. Using quotient rule, prove that if $y = \cot\theta$ then

$$\frac{dy}{dx} = -\text{cosec}^2\theta.$$

(4 marks)

b. Using a suitable substitution, show that

$$\int \tan\theta \; d\theta = \ln\sec\theta + C$$

(5 marks)
(total 9 marks)

11. Find the equation of the normal to the curve,

$$x = \sin(2y + \pi) \; at \; x = 1.$$

(5 marks)

12. The functions f & g are defined by

$$f: x \rightarrow 2x - 1, x \in \mathbb{R}$$

$$g: x \rightarrow \frac{x}{x + 1}, x \in \mathbb{R}, x > -1$$

a. Find $f^{-1}(x)$, stating its domain.

(3 marks)

b. Sketch the function $g(x)$.

(2 marks)

c. Find exact value of $gf\left(\frac{1}{4}\right)$.

(2 marks)

d. Find $g^{-1}(x)$, stating its domain.

(3 marks)
(total 10 marks)

Total for paper: 100 marks

End

GCE A Level Mathematics

Pure Mathematics 2

Potential Paper 2D

June 2021

Students must have Mathematical Formulae
and Statistical Tables, Calculator.

Calculator is allowed

Time allowed
2 hours
Total 100 marks

Write answers to 3 significant figures unless stated
otherwise

1. $f(x) = 4x^3 + 16x^2 + 9x - 9$

    i)   Using factor theorem or otherwise, show that $(2x - 1)$ is a factor of $f(x)$.

                                           (2 marks)

    ii)   Factorise $f(x)$ completely.

                                           (3 marks)

    iii)   Hence or otherwise, solve the equation

$$4(3^{3y}) + 16(3^{2y}) + 9(3^y) - 9 = 0$$

                                           (3 marks)
                                    (total 8 marks)

2. Calculate the area of triangle ABC.

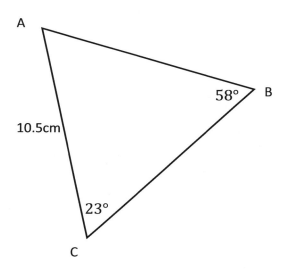

                                           (4 marks)

3. i) Given that $y = 3x^2(x^3 - 1)^4$, Show that

$$\frac{dy}{dx} = 6xf(x)(x^3 - 1)^3,$$ where $f(x)$ is a function to be found.

(4 marks)

ii) $x = \ln(cosec\ 3y), 0 < y < \frac{\pi}{6}$

Work out $\frac{dy}{dx}$ as a function of x.

(4 marks)
(total 8 marks)

4. a) Sketch the graphs $y = |3x|$ & $y = |x - a|, where\ a > 0$ on the same sketch.

(2 marks)

b) Solve $|3x| = |x - a|$
write your answers in terms of a.

(3 marks)

c) Solve $2 - x = 2|x + 1|$

(3 marks)
(total 8 marks)

5. i) Find using calculus, the x coordinate of the turning point of the curve with equation

$$y = e^{2x} \cos 3x , \qquad 0 < \theta < \frac{\pi}{3}$$

Give your answer to 4 decimal places.

(5 marks)

ii) Given that

$$x = \sin^2 3y , \qquad 0 < y < \frac{\pi}{4}$$

Find $\dfrac{dy}{dx}$ in terms of y. Write your answer in the form

$$\frac{dy}{dx} = p \, cosec \, (qy), 0 \leq y \leq \frac{\pi}{4}$$

Where p & q are constants to be found.

(5 marks)
(total 10 marks)

6. a. Prove that

$$\frac{\sin 2x \cos x}{cox \, 2x + 1} = \sin x$$

(3 marks)

b. Hence or otherwise, solve for $-\pi \leq x \leq \pi$

$$\frac{5 \sin 2x \cos x}{2 \, cox \, 2x + 2} = 3 \sin^2 x - 1$$

Give your answers to 3 significant figures.

(6 marks)
(total 9 marks)

7. $f(x) = (3 + kx)^{-4}$, $|kx| < 3, k$ *is a real constant.*

The expansion of $f(x)$ in ascending powers of x up to and including the x^2 term is

$$A + Bx + \frac{40x^2}{729}$$

Where A & B are real constants.

   i.   Write down the value of A

(2 mark)

   ii.   Work out the value of k.

(3 marks)

   iii.   Hence, write down the value of B.

(2 marks)

(total 7 marks)

8. A curve has parametric equations

$$x = \cot^2 \theta , \quad y = \cos \theta , \qquad 0 < \theta < \frac{\pi}{2}$$

   i.   Find an expression for $\frac{dy}{dx}$ in terms of θ.

(3 marks)

   ii.   Find an equation of the tangent to the curve at the point where $\theta = \frac{\pi}{4}$.

(5 marks)

   iii.   Find the cartesian equation of the curve in the form
$$y^2 = f(x)$$

(4 marks)

(total 12 marks)

9. (i) Given that $y > 0$, work out

$$\int \frac{2y - 3}{y(2y + 5)} \, dy$$

(6 marks)

(ii) Use the substitution $x = 9 \sin^2 \theta$ to work out the exact value of

$$\int_0^{\frac{9}{2}} \frac{\sqrt{x}}{\sqrt{9 - x}} \, dx$$

(8 marks)
(total 14 marks)

10. Prove that the sum of the first n terms of an arithmetic series is

$$S_n = \frac{n}{2}[2a + (n - 1)d]$$

(4 marks)

11. $f(x) = \frac{x}{2} + \sin x - 2\cos x$, where x is in radians.

    i)   Show that $f(x) = 0$ has a root in the interval $[0.9, 1]$.

                                                 (3 marks)

    ii)   Find an equation of the tangent to $y = f(x)$ at the point where it crosses the y-axis.

                                                 (4 marks)

    iii)   Find the values of constants a, b & c where $b > 0$ and $0 < c < \frac{\pi}{2}$, such that

$$f'(x) = a + b\cos(x - c)$$

                                                 (5 marks)

    iv)   Hence, find the x-coordinates of the stationary points of the curve $y = f(x)$ in the interval $0 \le x \le 2\pi$, giving your answers to 2 decimal places.

                                                 (4 marks)
                                   (total 16 marks)

Total for paper: 100 marks

End

GCE A Level Mathematics

Statistics and Mechanics

Potential Paper 3D

June 2021

Students must have Mathematical Formulae
and Statistical Tables, Calculator.

Calculator is allowed

Time allowed
2 hours
Total 100 marks

Write answers to 3 significant figures unless stated
otherwise

SECTION A: STATISTICS

1. The discrete random variable X can be only the values 3,4,5 & 6. For these values the probability distribution function is given below.

x	3	4	5	6
$P(X = x)$	$\dfrac{k}{8}$	$\dfrac{1}{8}$	$\dfrac{1}{2k}$	$\dfrac{1}{k}$

Where k is a positive integer with $k < 4$

i) Show that $k = 3$

(3 marks)

ii) Work out $P(X \geq 4)$

(3 marks)

iii) Work out $E(X)$

(3 marks)

iv) Work out $E(X^2)$

(3 marks)

v) Hence, work out $Var(4x - 1)$

(3 marks)

vi) Work out $E(3x + 2)$

(2 marks)
(total 17marks)

2. Information about some salaries are given below.

Salaries (£)	Number of people (f)
1-1000	40
1001-1500	110
1501-2500	170
2501-4000	105
4001-5000	78
5001-10000	12

i)   Work out the mean salary

(3 marks)

ii)   Work out the median, lower quartile & upper quartile salaries

(5 marks)

iii)   Hence, write down the interquartile range

(1 marks)

iv)   Represent above data on a histogram

(3 marks)

v)   Calculate the standard deviation of the above salaries

(2 marks)
(total 14 marks)

3. 70% of adults weigh over 65kg and 90% of adults weigh over 55kg in a city. The weights of adults are said to be normally distributed with mean μ & standard deviation σ.

    i)   Find μ & σ.

(5 marks)

    ii)   Work out the percentage of adults between 70kg and 75kg.

(4 marks)
(total 9 marks)

4. In a particular town, the chance of a road accident is 0.6 per day. The council had recorded only 6 accidents in the last 30 days.

A councillor claims that the mean number of accidents per day have changed.

Test this claim at 5% significance level clearly stating your hypothesis.

(5 marks)

5. 64% of a village population have a dishwasher and 90% have a washing machine. Given that the two events are independent.

    i)   Calculate the percentage of population who have both appliances.

(2 marks)

    ii)   Given that a person has a washing machine, work out the percentage of this person also having a dishwasher.

(3 marks)
(total 5 marks)

Total for Section A is 50 marks

SECTION B: MECHANICS

6. A block of mass 10kg is sliding down a smooth surface inclined at 20° to the horizontal as shown below. The block is 25m away from the bottom of the slope when it is released from rest.

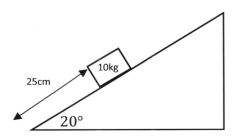

i)   Calculate the time taken for the block to reach the bottom of the slope.

                                                                  (5 marks)

A horizontal force F is applied to the block so that, it now accelerates $2ms^{-1}$ upwards.

ii)   Work out the value of F.

                                                                  (6 marks)

iii)   Calculate the normal reaction force, during this this upwards acceleration.

                                                                  (5 marks)
                                                            (total 16 marks)

7. A uniform rod AB of mass 15kg and length 10m is resting in equilibrium on two supports at A and C. Point C is 2m from B.

    i)   Calculate the reaction force at C.

                                             (5 marks)

    A boy of mass 45kg stands on the rod at the point D. The beam remains in equilibrium. The reaction force at A is now 1.5 times the reaction force at C.

    ii)   Calculate the distance AD.

                                             (6 marks)

    The boy now sits at B.

    iii)   The beam is about to tilt about C. Calculate the reaction force at C.

                                             (5 marks)
                                   (total 16 marks)

8. A particle P of mass 3kg is moving under the action of a constant force of F Newtons. When t = 0, P has velocity $(i + 5j)ms^{-1}$ and at time t = 4 s, it has velocity $(12i - 8j)ms^{-1}$.

    i)   Find the acceleration of P in terms of i and j

    (5 marks)

    ii)   Work out the magnitude of F

    (4 marks)
    (total 9 marks)

9. An aircraft lands with a velocity of Ums^{-1} and comes to rest after 24 seconds on a runway which is 800m long.

    i)   Calculate the deceleration of the aircraft.

    (5 marks)

    ii)   Work out the value of U.

    (4 marks)
    (total 9 marks)

Total for Section B is 50 marks

Total for paper: 100 marks

End

Answers

Paper 1A

1. Correct proof

2. min $\left(\frac{2}{3}, \frac{37}{27}\right)$, max $\left(\frac{1}{2}, \frac{11}{8}\right)$

3. $P = 11.5cm, A = 7.5cm^2$

4.
i) $(x + 1)(3x - 4)(2x + 1)$
    ii) *correct sketch*

5. i) $3 - \frac{1}{3}x - \frac{1}{54}x^2 + \cdots$
    ii) 2.997

6. i) $k = 20$,
    ii) *correct sktech*

7. $\theta =$
    $0.62, 3.76, -2.52, -5.66$

8. a) £24550, b) £7426.12,

    c) 7 years, d) £550,
    e) sketch & the car will
always be worth at least £550.

9. i) correct sketch
    ii)a)$y = -\frac{1}{3}x$
    b)$x = 0$ or $x = \frac{10}{3}$
    iii) $\frac{500}{81}$

10. i) correct sketch,
    ii)$x = \frac{2}{3}$ & $x = \frac{2}{5}$

11. 60

12. $26x - 11y - 59 = 0.$

13. Correct proof.

14. a) correct sketch
b) correct proof, c) 11.9

Paper 2A

1.i) $f^{-1}(x) = \frac{2x+4}{x-3}$, ii) $f^{-1}(x) > 2$,
iii) $x > 3$

2. $x : y = 1 : 1$

3. i. correct proof, ii. $0.922cm^2$

4. i) $\frac{dy}{dx} = -\sin\theta$,
ii) $4y + 4x - 7 = 0$

5. a. $f(1) = 0$, b. $(x - 1)(x^2 - 3)$,
c. $\frac{\pi}{4}, \frac{-3\pi}{4}, \frac{\pi}{6}, \frac{-5\pi}{6}, \frac{-\pi}{6}, \frac{5\pi}{6}$

6. a. $\theta = 0.983^c$ & 4.12^c
b. $x = 60°, 180°$ & $300°$

7. i) 7,971,600, ii) $\frac{8}{9}$

8. proof using quotient rule.

9. i) $R = 25, \alpha = 0.284$,
ii)0.851^c & 6.02^c,
iii). a. 15625, b. 1.85^c

10. i) $2\tan^3 x + c$,
ii) $\frac{8}{3}\ln 2 - \frac{7}{9}$

11. i) $\frac{dy}{dx} = \frac{ye^x - e^y}{xe^y - e^x}$, ii) correct proof

12. a) $\frac{1}{2(3+y)} + \frac{1}{2(3-y)}$,
b) $\sin^2 x = \frac{1}{8}\left(\frac{3+y}{3-y}\right)$

13. correct proof.

Paper 3A

1. i) $\mu = 32.6, \sigma = 35.2$
ii) $a = 14.1$

2. i) tree diagram
ii)$\frac{19}{66}$, iii) $\frac{16}{47}$

3. i) $\bar{x} = 156, \sigma = 8.09$
ii) There is insufficient evidence to
reject H_0. Therefore managers claim is
incorrect.

4. i) $\bar{x} = 29.7, \sigma = 8.5$
ii) $Q_2 = 29, IQR = 15$
iii) correct proof

5. i) a. 0.0667, b. 0.0739
ii) Reject H_0, Inspectors claim is
correct.

6. $i)$ $5.47ms^{-2}, ii)$ $34.6N, iii)$ $3.12s$

7. $i)$ $35.04m, ii)$ $22.8m, iii)$ $25.6ms^{-1}$

8.
$i)$ *diagram with all forces, $ii)$* 0.289

9. $i)$ $212.3N, ii)$ $0.38m$

10. $a)$ $R = 66.6N, b)$ $a = 2.5ms^{-2}$

Paper 1B

1. $(x - 2)(4x + 3)(3x - 1)$

2. a) $5 - \sqrt{2}$, b) proof

3. a. $f'(x) = \frac{(x-3)(3x+1)}{(3x-1)^2}$

   b. $-\frac{1}{3} < x < 3$

4. i) $n = -\frac{1}{3}, a = -2$
   ii) proof

5. a) $p = 3, q-= -2, r = -5$
   b) sketch
   c) translation by vector $\binom{1}{7}$
   d) $h(x) \geq -2$

6. i) $x = 1$
   ii) $\theta = \pi, \frac{\pi}{6}, \frac{5\pi}{6}$

7. i) $x = \ln 16$ & $x = \ln 3$
   ii) proof

8. i) sketch
   ii) $y - 11x + 39 = 0$
   iii) $\frac{59}{132}$

9. i. $a = \frac{2}{3}$
   ii. $x = 7.45$

10. i) sketch
    ii) $x = 2$ & $\frac{3-\sqrt{41}}{2}$

11. a. 63214
    b. 23 years and 164 days

12. i) $(0,0), (2,\frac{4}{e^2})$
    ii) $ey + e^2x - 1 - e^2 = 0$

13. a) $\frac{1}{3x-1} + \frac{2}{(1+x)^2}$
    b) $\frac{1}{3}\ln\left(\frac{5}{2}\right) + \frac{1}{3}$

Paper 2B

1. i) $k = \frac{17\sqrt{10}}{10}$ ii) $n = 72$

2. 2.321

3. i) $r = 8cm, \theta = 1.25$
   ii) $9.63cm^2$

4. i) proof
   ii) $\frac{4x^2}{(x^2+1)^2} + \frac{4y^2}{(y^2+1)^2} = 1$
   iii) $\frac{dy}{dx} = -\frac{y^2+1}{x^2+1}$

5. i) $-\frac{1}{3}\ln(3 - x^3) + c$
   ii) $\frac{1}{4}(3 - \sqrt{2})$

6. a) $f(x) > 0$
   b) $f^{-1}(x) = 2 + \ln\left(\frac{x}{2}\right)$
      domain $x > 0$
   c) $\frac{18}{e^2} - 1$
   d) $x = \frac{1}{3}(2e^3 + 1)$

7. i) $\frac{1}{2}x^2 = 441(4t - \frac{t^2}{2})$
   ii) verify

8. a. $r = 2$
   b. proof
   c. 201.29

9. i. $\overrightarrow{AB} = \begin{smallmatrix} 1 \\ 5 \\ -4 \end{smallmatrix}, \overrightarrow{AC} = \begin{smallmatrix} 3 \\ -1 \\ 0 \end{smallmatrix}$
   $\overrightarrow{BC} = \begin{smallmatrix} 2 \\ -6 \\ 4 \end{smallmatrix}$
   ii. $\sqrt{42}, \sqrt{10}$ & $2\sqrt{14}$
   iii. $95.6°$
   iv. 10.2

10. i. separate sketches with
    a. $(b, 0)$ & $(0, a)$
    b. $\left(\frac{a}{2}, 0\right)$ & $(0, 3b)$
    ii. $a = -3$ & $b = -1$
    iii. $f^{-1}(x) = (1 - x)^2 - 4$
    domain $x \leq 1$

11. a. $f(2.2) = 0.02004 > 0$
    $f(2.3) = -0.08720 < 0$
    Therefore, there is a
    root in the interval
    b. $f'(x) = \cos x - \frac{1}{x}$
    c. 2.219

12. a. $R = 25, \alpha = 0.284$
    b. $\theta = 0.696^c, 3.014^c$
    c. max $h(\theta) = 3$
    d. $x = 0.618$

Paper 3B

1. a. $k + 4k + 9K = 1,$
   $k = \frac{1}{14}$
   b. $\frac{13}{14}$
   c. $\frac{18}{7} = 2.57$
   d. $var(x) = 7,$
   $var(2 - x) = 7$

2. i. 71.2
   ii. 67.3
   iii. 12.0
   iv. 30

3. a. $\frac{1}{5}$
   b. $\frac{4}{5}$
   c. $\frac{3}{10}$

4. a. 0.3436
   b.
   $p(x \leq 2) = 0.1024 > 0.1$
   Accept H_0
   the claim in the advert is correct.

5. i. Mean=47 years
   standard deviation 51
   years
   ii) 48

6. a. 06769
   b. 0.6700
   c. 1.02%

7. 340m
8. a.125m
   b.3405m/s

9. i. $4.53ms^{-2}$
   ii. 3.15s

10. 1.29m
11. a. $4.9ms^{-2}$
    b. $4.95ms^{-1}$
    c. 6.25m

Paper 1C

1. i. £56
   ii. 45 days

2. i. $C(2,7)$
   ii. $(x-1)^2 + (y-7)^2 = 5$
   ii. $PQ = \frac{21\sqrt{5}}{2}$

3. i. $2x - 8$
   ii. $\frac{x^3}{3} - 4x^2 + 3x + c$
   iii. $12y - x - 278 = 0$

4. a. $A = \frac{2}{5}$, $B = \frac{2}{15}$, $C = \frac{2}{3}$
   b. $\frac{2}{15}(9 - 11x + 13x^2)$

5. i. $\overrightarrow{AB} = \begin{matrix}4\\-2\\3\end{matrix}$
   $\overrightarrow{AC} = \begin{matrix}8\\-4\\3\end{matrix}$
   ii. $15.3°$
   iii. 6.70

6. $r = 3cm$, Area $2.26cm^2$

7. Proof
8. i. 15000
   ii. 59.2 days
9. i. $-4cosec^2 2y cot 2y$
   ii. proof
   iii. $\frac{6x-4}{16x^2(x-1)^{\frac{3}{2}}}$
10. i. $y = 1 - \sqrt{1 - (2-x)^2}$
    ii. $\frac{1}{2}$
    iii. $y + 2x + 3 = 0$

11. i. proof
    ii. $c = -\frac{17}{8}$
    iii. $\frac{x+5}{(x+1)^3}$

12. i. $R = \sqrt{58}$ $\alpha = 0.405$
    ii. proof
    iii. $0.225, 2.51$

Paper 2C

1. $16x^{\frac{7}{2}} + c$

2. $x = 1 \ or \ 10$

3. $\frac{1}{27} + \frac{4x}{27} + \frac{32x^2}{81} + \frac{640x^3}{729}$

4. i) $f(x) > 0$
   ii) $3 + \ln(\frac{x}{2}), x > 0$
   iii) $8e^{-3}$
   iv) $\frac{1+2e^2}{3}$

5. i. $\frac{x^2+2x+4}{3x-2}$
   ii. $x = \ln 2 \ or \ \ln\frac{2}{3}$

6. i. $-6 \cos x \sin x - 2cosec\ 2x \cot 2x$
   ii. $2(1 + \frac{1}{x})(x + \ln 3x)$
   iii. $\frac{-14x-14}{(x-2)^2}$

7. i. 50,000
   ii. 3.39
   iii. 120,000

8. i. proof
   ii. $\theta = 0.190, 1.76, -1.38, -2.95$

9. i. sketch
   ii. $x = \frac{1}{2} \& -1$
   iii. $x = 0 \& -4$

10. i. $\frac{1}{6}\ln(2e-1)$
    ii. $\frac{1}{\sqrt{3}} - \frac{\pi}{6}$
    iii. $\frac{\sqrt{3}}{2} + \frac{\pi}{6}$

11. i. $4\pi r^2$
    ii. $\frac{125}{\pi r^2(4t+1)^2}$
    iii. $v = 125 - \frac{125}{4t+1}$
    iv. a. r = 3.08cm
    b. 0.0025cm/s

Paper 3C

1. i. Normal Distribution $X\sim N(2.1, 0.5^2)$
   ii. 3.59%
   iii. 0.673
   iv. 2.52

2. i. table
   ii. $\sigma_x = 2.21$ $\sigma_y = 0.189$
   iii. $\sigma_T = 22.1$ $\sigma_M = 18.9$
   iv. $M = 4.4T - 250$
   v. 410
   vi. Unrealistic as it is extrapolation

3. i. 0.14
   ii. 0.22
   iii. $\frac{2}{3}$
   iv. $\frac{18}{25}$

4. i) Histogram
   ii) mean=164.38cm Standard Deviation 17.99cm

5. i) 0.5856
   ii) n large & p close to 0.5
   iii) Reject H_0, schools' claim is wrong

6. i. 1000N
   ii. 1375N

7. 267.3m
8. i. .31.6m/s
   ii. 111.9m
   iii. 6.39seconds

9. i. 73.5N
   ii. 5.36m
   iii. 1021N

10. i. max P=52N min P = 11.6N
    ii. $6.87ms^{-2}$

Paper 1D

1. i. *sketch showing* $(5, -28), (0, -3)$
   ii. $6y - x + 116 = 0$
   iii. $\frac{3364}{3} unit^2$
2. proof
3. i. $(x - 6)^2 + (y - 5)^2 = 20$
   ii. $y - 2x - 3 = 0$
4. i) $\frac{1}{10}\ln(\frac{5e+3}{8})$
   ii. $\sqrt{3} - 1 - \frac{\pi}{12}$
   iii. $\frac{3\pi}{2}$
5. proof

6. i. $\log R = \log a + kt$
   ii. graph
   iii. $R = 15.8 \times 10^{0.16}$
   iv. £15800
   v. 12 years

7. i. $r = \frac{1}{3} \& a = 492075$
   ii. 12.5

8. $x < 4 \ \& \ x > \frac{2}{3}$
9. i. $\frac{3\sqrt{3}}{4}$
   ii. $(\frac{4}{3}\sqrt{\frac{3}{5}}, 9\sqrt{\frac{3}{5}})$
   iii. $4y - 3\sqrt{3} - 18 = 0$

10. a. proof
    b. proof
11. $y = -\frac{\pi}{4}$
12. a. $f^{-1}(x) = \frac{x+1}{2}$, $x \in \mathbb{R}$
    b. sketch
    c. $gf\left(\frac{1}{4}\right) = -1$
    d. $g^{-1}(x) = \frac{x}{x-1}$, $x < 1$

Paper 2D

1. i. proof
   ii.
   $$(2x - 1)(x + 3)(2x + 3)$$
   iii. $y = -\frac{\ln 2}{\ln 3}$
2. $50.4 cm^2$

3. i. proof
   ii. $\frac{dy}{dx} = \frac{-1}{3(e^{2x}-1)^{\frac{1}{2}}}$
4. a. sketches
   b. $x = \frac{a}{2} \& \frac{a}{4}$
   c. $x = 0 \& -4$
5. i. 0.1960
   ii. $p = \frac{1}{3}, q = 6$
6. a. proof
   b. $-0.300 \ \& \ -2.842$
7. i. $A = \frac{1}{81}$
   ii. $k = 2$
   iii. $B = -\frac{8}{243}$
8. i. $\frac{\sin^4 \theta}{2 \cos \theta}$
   ii.
   $$8y - \sqrt{2}x - 3\sqrt{2} = 0$$
   iii. $y^2 = \frac{x}{x-1}$
9. i. $\frac{1}{5}\ln(\frac{(2y+5)^8}{y^3}) + c$
   ii. $\frac{9}{4}(\pi - 2)$
10. proof
11. i. change of sign shown
    ii. $2y - 3x + 4 = 0$
    iii. $a = \frac{1}{2}$, $b = \sqrt{5}$, $c = 1.11$
    iv. 2.91 & 5.60

Paper 3D

1. i. proof
   ii. $\frac{5}{8}$
   iii. 4.46
   iv. 21.54
   v. 6.66
   vi. 15.4

2. i. £2484.95
   ii. median £2132.85
   Q1=£1403.91
   Q2=£3446.93
   iii. £2043.02
   iv. histogram
   v. £1424.70

3. i. S.D=13.2kg
   Mean=71.9kg
   ii. 14.67%

4. Reject H_0
   Mean has decreased

5. i. 0.576
   ii. 64%

6. i. 3.85seconds
   ii. 56.95N
   iii. 111.6N

7. i. 91.9N
   ii. 2.6m
   iii. 367.5N

8. i. $4.26 ms^{-2}$
   ii. 12.78N

9. i. $-2.78 ms^{-2}$
   ii. $66.7 ms^{-1}$

Printed in Great Britain
by Amazon

58225499R00070